Cool Careers in
INFORMATION SCIENCES

Sally Ride
Science

CONTENTS

Katrin

Michael

Cristina

Zoran

Jeannette

Brewster

Hiroshi

Jenna

Anthony

Patricia

Atul

Marissa

What Do You Want to Be?

Is working with digital information one of your goals?

The good news is that there are many different paths leading there. The people who work in information science come from many different backgrounds. They include computer scientists, biologists, engineers, medical doctors, science writers, social scientists, inventors, and many more.

It's never too soon to think about what you want to be. You probably have lots of things that you like to do—maybe you like doing experiments or drawing pictures. Or maybe you like working with numbers or writing stories.

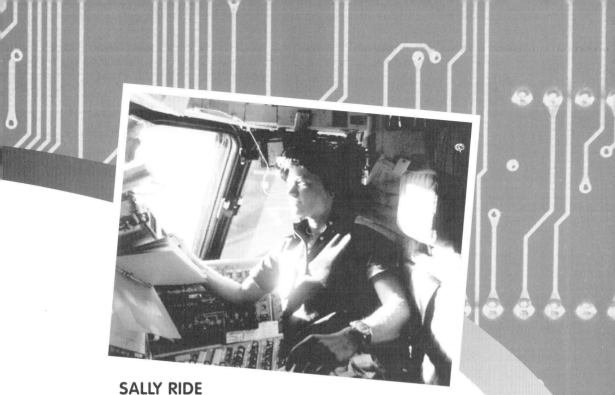

SALLY RIDE
First American Woman in Space

The women and men you're about to meet found their careers by doing what they love. As you read this book and do the activities, think about what you like doing. Then follow your interests, and see where they take you. You just might find your career, too.

Reach for the stars!

Sally K Ride

U Can, 2

Even students can get into the act! "Imagine your teacher gets a text, 'We need 10 students 2 help @ a tree planting—RU interested? Text *Yes* if U R and we'll tell U where 2 go.' That's just one example," Katrin says.

■ KATRIN VERCLAS

MobileActive.org

Dial "B" for Better

Katrin Verclas decided at 16 she wanted to make the world a better place. At first, Katrin thought she'd become a doctor or nurse—but she wanted to do more than just help one person at a time. Katrin wanted to reach out and touch the lives of lots of people at once. So she picked up a tool that connects most of the world—the mobile phone—and got to work.

Anywhere, Anytime

Katrin now runs an online community of people who use mobile phones in positive ways. One way is by helping sick women in Mexico connect with each other. They share tips about doctors and medications by texting each other. Another mobile phone project reaches people in countries with restrictions on news and speech, such as Zimbabwe. It encourages them to use their phones to report their own news stories. "My job is to be the glue among these projects—to write about them, collect and tell their stories, find money for them, and bring people together," says Katrin.

Most children outfox their parents when using mobile phones. Not Katrin's! "The day my kids outsmart me, I should look for a new job!"

An activist . . .

works to inform people and bring about change—usually in ways that make the world a better place. Katrin writes about ways people can share useful information by mobile phone. Other **activists**

- ☐ write letters to newspapers to draw attention to issues.
- ☐ help create new laws to protect plants and animals.
- ☐ form support groups for people with disabilities.

Phone Home

The field trip is cancelled. Your teacher asks you to get the word out—quickly. You decide to call two classmates and ask each of them to call two more. How many steps—sets of calls—will it take to contact 30 classmates? Write an equation to show your answer.

Air Time

Katrin and her two children like to experiment with new mobile applications. One application finds simple volunteer projects you can do by phone.

Brainstorm with a partner about a project in your community that volunteers could do using their mobile phones.

Night Times

Before bed each night, Daniel, Monica, Mia, and David each text "sweet dreams" to their parents in California. These four grown kids all live in different states—Oregon, Colorado, Missouri, and Virginia. That means they also live in different time zones. But, Mom and Dad receive all four messages exactly at 9 P.M. California time. In which state does each of the four children live if

- Daniel goes to bed earliest?
- Mia is the second to last to go to bed?
- Mia goes to bed earlier than Monica?
- Monica goes to bed later than David?

Ring Around the World

Of the nearly 7 billion people in the world, a whopping 4 billion of us have mobile phones!

Check out your answers on page 36.

MICHAEL QUINN
Seattle University

The Right Stuff

How much control should your parents have over which Web sites you can visit? Is it wrong to watch a movie if you know it's a pirated copy? Should cell phones be permitted at school? Mike Quinn spends a lot of time thinking about these kinds of tricky questions that come up as new technologies take hold. As an ethics professor, he says it's not his job to tell people what to do. Mike just hopes his students will think more about the implications of decisions they make.

Responsibil-IT

Imagine you're a software engineer. For months, you've been developing a program that will store patients' medical records. You notice a flaw. It means that people's private information could wind up exposed. The project is due tomorrow, however, and fixing the problem means missing the deadline. Your managers are really pushing for on-time delivery. Would you be able to stand up to your bosses to protect people's private information? "Like doctors or lawyers, people working in information technology make decisions that affect other people," Quinn says. "And sometimes doing the right thing takes guts."

"The introduction of computers has created some ethical problems that haven't been thought of before."

Producing a Dream

In high school, Mike dreamed of making movies. But the equipment was too expensive. Now that the technology has progressed, he makes music videos of his son's band. "Computers and software have made it possible for me to fulfill one of my dreams," he says.

"When I was a kid, I wondered how boats could sail upwind. When I sailed, I quickly learned that the sail is just a big wing."

A computer ethics expert . . .

analyzes and tries to answer philosophical questions about moral behavior. Mike uses traditional ethical theories to consider issues raised by new technologies. Other **ethics experts**

- ☐ wrestle with questions having to do with advances in medicine.
- ☐ consider how companies should treat their customers.
- ☐ debate policy issues that determine how we deal with many sensitive topics.

Video Villains

File-sharing sites on the Internet are a popular way of downloading movies, music, and TV shows. However, many of these sites are illegal in the U.S. Take a stand on the ethics of this issue. Should file-sharing sites remain illegal or be legalized?

- Divide into teams.
- Each team researches its position. Teams come up with several reasons and examples to support their arguments.
- Hold a class debate.
- Afterwards, discuss with the class how the debate might have affected your thoughts about the issue.

Did anything change your mind? If so, what was it?

Flying Through Time

Construct a time line showing how a nifty technology, such as the telephone, has changed over time. Be sure to include

- the year the device was invented.
- date(s) the device was significantly improved, and how it was improved.
- photos and/or drawings with captions that describe the upgrades.

Present your technology and time line to the class. Predict what the next generation of the device might be like.

About You

When you make an ethical decision, you make a judgment. In your About Me Journal, write about a time you had to make a difficult decision. How did you decide what to do? Describe how ethics influenced your judgment. Looking back, do you think your decision was the right one? Why?

CRISTINA VIDEIRA LOPES

University of California, Irvine

Cheap 'n Slow

CU@practice@4. Text messages zip between cell phones on radio waves traveling at the speed of light. So why would Crista Lopes want to slow down wireless communication to the speed of sound—about 875,000 times *slower*? It saves money. Sound waves can be a cheaper way of swapping data between electronic devices, such as laptop computers. With sound, you don't need a radio transmitter or radio receiver—just a microphone, speakers, and Crista's nifty software.

> Crista loves programming because it allows her to turn thought into action. "You just pop up with an idea and make it happen."

Good to Hear

Buzzzz. Cracccccckle. Hummm. Two laptops swapping phone numbers or addresses using sound waves would sound really annoying. So Crista tailors her software to encode, or hide, any message in pleasant and natural sounds— like a bird or chirping cricket. "They sound very natural," Crista says. The first laptop encodes the message and then broadcasts it using ordinary speakers. *Chirp-chirp!* The second laptop picks up the message with its microphone. It uses Crista's software to decode and display the hidden message. You hear just the nice chirping—but not the encoded message. "You don't even notice it!" Crista says.

Crista and her students view a virtual world to demo a futuristic rapid transit system.

Out Loud

Crista loves computer programming as much as she loves to express herself out loud—she used to sing with the San Francisco Symphony Chorus. She now puts on opera concerts at home for her two children.

A computer programmer . . .

uses computer languages to write software that runs on computers. Crista writes programs that guide computers to exchange information using sound waves. Other **computer programmers**

- ☐ develop video games.
- ☐ produce 3-D graphics for Web sites.
- ☐ write instructions for robotic spacecraft.
- ☐ turn computers into musical instruments.

Sound Fast?

Radio waves travel at the speed of light—a blazing 300,000 kilometers (186,000 miles) per second. At that speed, how long would it take a radio message to travel from mission control on Earth to a crew of astronauts on the Moon, a distance of 384,400 kilometers (238,855 miles)?

Sound waves travel at much pokier speeds—at about 343 *meters* (1,125 *feet*) per second. Now imagine the launch of a Moon rocket from Kennedy Space Center in Florida. *Vroom!* How far would that sound travel in the amount of time it takes to travel from Earth to the Moon at the speed of light?

Flash . . . Bang!

Have you seen a bolt of lightning during a storm, then a few seconds later heard the clap of thunder it produced? Whoa, that sounded close! Next time there's lightning, stay safely inside, then calculate how close it is.

- After the lightning strikes, immediately start counting— *one* one thousand, *two* one thousand, etc.—until you hear the thunder clap.
- Say you count to four. How far away, in kilometers, is the lightning? Remember the speed of sound? That means for each second you count, a thunder clap travels 343 meters (1,125 feet).
- How many miles away is the lightning? One mile is 5,280 feet.

La, La, La, Laugh

Knock-knock.
　　Who's there?
Opera.
　　Opera who?
Opera-tunity—and you thought opportunity knocked only once!

ZORAN POPOVIĆ

University of Washington

Not-So-Smooth Moves

Can you spot a computer-generated fake? If it's an animated character in a video game or movie, Zoran Popović says, "Yes!" The character may appear realistic—until it moves. "One of the big problems in computer graphics is how to create virtual characters whose motions aren't stiff or artificial," Zoran says. So Zoran is looking for solutions. He studies how humans—and other animals—move their bones, muscles, and skin. Then he creates software so computer animators can recreate that motion more smoothly and naturally.

"It's so exciting to see the impact new discoveries can quickly have."

Smooth Moves

Zoran's lifelike animated characters walk, run, and dance in realistic ways—their moves are less jerky or clunky. That's not all Zoran gets moving—he also wrote computer code that adds more *zoom* to racecars! ESPN uses the software to show the flow of air around stock cars during NASCAR races. The zoom in Zoran's career is what he loves about it—computer science moves fast and changes fast. In fact, studying computer science is one of the reasons he chose to remain in the U.S. after his year as a high school exchange student from the former Yugoslavia.

Dino-mation

Even extinct creatures can get the Zoran treatment. Take the skeleton of an ancient dinosaur. You could use Zoran's computer graphics software to analyze all the possible ways dino bones work together. "That way, you can actually say what was the most likely way the dinosaur moved!" Zoran says. Can you say, "run, ty-*run*-osaurus, run?"

A computer scientist . . .

designs computers and software—and new ways to use both. Zoran designs software used to create virtual characters that move in natural ways. Other **computer scientists**

- ◘ design networks that allow people to send data between computers.
- ◘ create more realistic graphics for video games and movies.
- ◘ program robots that can mimic human abilities.
- ◘ invent better ways of searching the Web.

Nibble Kibble

Zoran says, "My next project is a computer game that teaches students fractions—just by playing!" In one of the activities, players must keep their virtual pets happily fed. Imagine your virtual pet eats 1½ scoops of virtual pet food a day. How many scoops does it eat in

1. one week?
2. a 30-day month?
3. one year?

Is It 4 U?

Computer scientists like Zoran use computer animation to teach and learn. They might

- experiment with artificial limbs on virtual patients.
- create computer games that solve scientific puzzles.
- design ways for robots to move more like humans.
- catalog ways the human face can express emotion.

Discuss with a partner what parts of Zoran's work you would like, and why.

Real Deal

One day, Zoran imagines software will let us give virtual characters on the computer screen the same expressions and gestures as real people. Brainstorm with a partner and list different ways you would animate a computer character who has just

- seen a long-lost friend.
- tasted unexpectedly spicy food.
- heard the floorboards creak late at night.
- missed the last bus home.

Check out your answers on page 36.

JEANNETTE WING

Carnegie Mellon University

Tough Stuff

How do you keep the private information on computers safe from hackers and other snoops? That's the problem Jeannette Wing is trying to solve. It's a lot like locking your diary, putting it in a box and taping the box shut, and then putting the box in a safe. "When we design computer systems to be secure, we think in the same way. We add layers and layers of security, so if attackers make it through one layer, they are stopped by the next layer," Jeannette says.

Kick It

Jeannette is also an expert in defense. She even has a black belt in Tang Soo Do, a Korean martial art. When she's not doing karate, Jeannette practices ballet or yoga. "I'm interested in all those things that stretch you," she says.

Real Memory

Growing up, Jeannette never knew what to say when she was asked what her father did. "So my mom taught me to repeat, 'My father is a professor of electrical engineering.' I memorized it but never knew what it meant until high school!" Jeannette says. That's when she learned her father used math and science to make things work. In college, Jeannette wound up changing her major from electrical engineering—but only after checking with her dad to make sure it was okay. "I didn't want to switch if computer science was just a fad!"

A computer security expert . . .

makes sure computers and the information they hold stay safe. Jeannette researches how to keep intruders from gaining access to computer systems and the information they contain. Other **computer security experts**

- ◘ keep viruses from infecting computers.
- ◘ make passwords easy to remember but hard to guess.
- ◘ track down stolen data on the Internet.

Learning your QWE's

You can send a secure message to a friend by creating your own code. Here are three different ways.

- Shift each letter of the alphabet two letters the right, so $A = C$, $B = D$, and so on.
- Assign numbers to each letter, starting from the end of the alphabet, so $Z = 1$, $Y = 2$, and so on.
- Assign alphabet letters in order of the letters on a computer keyboard starting with the top row of letters, so $Q = A$, $W = B$, $E = C$, $R = D$, and so on.

Use one of these or create your own code. Then write a one-sentence message about something you did this week. Give it to a classmate to decode. You might need to pass on some clues about your code.

Low to High

We're surrounded by security—some high-tech and some low-tech. Jeannette's computer security is definitely high-tech. A padlock is low-tech. With a team,

- make a list of all the low-tech security around you—for example, at school, in stores, and at home.
- brainstorm how each could be computerized.
- create a chart called "From Low- to High-Tech," and list your ideas for updating.

Compare your chart with other teams.

A Little Help from My Friends

Jeannette checked with her dad before changing her college major. Who helps you make big decisions? In your About Me Journal, write an informative paragraph that describes the person and how he or she helps you.

Also include what you've learned from the person that could help you make big decisions in the future.

BREWSTER KAHLE

Internet Archive

Quick to Change

Brewster Kahle set a huge goal for himself when he was in college. "I wanted to make all knowledge available to everyone, everywhere—I wanted to put everything online." Wow, we're talking zillions of books, newspapers, Web pages, movies, and much, much more. Brewster started building the vast digital library called the Internet Archive in 1996. It gives anyone, anywhere in the world, access at any time to all of its information—for free.

That Library is 'brary Big

So far, the Internet Archive holds 150 billion (yes, billion!) Web pages that were created by schools, businesses, news organizations, individuals, and others. It also holds more than 1 million books, 250,000 sound recordings, and 135,000 videos. And it's still growing, fast. You can't find everything in the Internet Archive—yet. But every year Brewster and his team add more and more and more. "I never have to wonder what I'll do next!" Brewster says.

Books on Wheels

The Internet Archive also has a real bookmobile. Boys and girls can download books from its online collection and then print and bind them right on the spot. "You can make books available even to people who don't have a computer," Brewster says.

The Internet Archive Bookmobile could be coming to your town.

A digital librarian . . .

collects, organizes, digitizes, and preserves digital information and helps people find it. Brewster founded an Internet library that contains billions of Web pages, books, films, and sound recordings. Other **digital librarians**

- ◘ teach people how to use the Internet.
- ◘ scan and digitize books.
- ◘ archive Internet news stories on important events.
- ◘ collect links to useful Web sites.

What's the Big Idea?

Brewster says the Internet Archive gave direction to his life. Create a list of "big ideas" you would like to tackle. Choose one. Then write a proposal to a company that could fund your project. Here are some things to include.

- • What is your idea?
- • Why it is important?
- • How will it help others?
- • What will you need to get started?

Present your proposal to several classmates who will act like a company review board. Would they write you a check? Why or why not? Use their feedback to help take your idea to the next step.

To Print or Not to Print?

That is the question—for your class to debate! In the future, should books still be printed or made available only online? Take a class poll and record the results. Next, divide into teams to research newspaper or magazine articles on the topic, and then hold a debate.

> Printed books are no longer needed.
> *vs.*
> Printed books will always be needed.

Afterward, take another class poll. What percentage of the class was persuaded to change their minds?

Is It 4 U?

As a digital librarian, Brewster enjoys

- • working with a team.
- • designing computer systems to catalog information.
- • organizing a big project that could last for many years.

What parts of Brewster's work interest you? Discuss with a partner what skills you have that would make you a good digital librarian.

HIROSHI ISHII
Massachusetts Institute of Technology

Touchy Bits

When Hiroshi Ishii works with computers, he's a real hands-on guy—even if he never touches a single keyboard or mouse. Instead of typing and clicking, you're more likely to see Hiroshi sculpting, brushing, or even digging with his hands. That's because Hiroshi invents new ways to artistically *and physically* interact with the digital information stored on computers and other devices.

Paint of Coat?

Hiroshi's imaginative inventions blur the boundary between the real world we live in and the virtual world of computers. One invention builds a teeny digital camera into a computer-linked brush such as the one below. Just hold up the soft brush bristles to, say, a pile of marbles or a Dalmatian's hair. The camera inside the brush captures colors, textures, and any movement—and stores it. "The brush lets you make your own ink from the world," Hiroshi says. Next, move the brush over a special computer screen. Each brushstroke will come out with the "ink" you've stored. On the screen you'll see marble colors and shapes, or furry-looking black and white spots!

The Hole Story

Mixing the digital and physical worlds has a long history. When Hiroshi was growing up in Japan, his father, a computer programmer, used to bring home punch cards. Early computers stored digital data on these cards by punching in patterns of holes. Hiroshi loved to play with them. "The physical representation expressed the program," Hiroshi says.

—— Camera

—— Touch sensors

A digital visionary . . .

imagines the role computers will play in the future—and then creates that vision. Hiroshi invents ways to physically handle the information stored digitally on computers. Other **digital visionaries**

- ◘ make low-cost computers available to developing nations.
- ◘ add computers to fridges, dryers, and other "smart" appliances.
- ◘ turn cell phones into mobile computers.

News That Glows

Another of Hiroshi's digital inventions is a frosted-glass globe that glows in different color alerts depending on the information displayed. For example, the globe glows green when the stock market prices are rising and red when they're falling.

- Come up with your own application for a glowing information globe.
- What would you want the globe to report?
- What colors would indicate which information?

Create and illustrate an advertisement for your globe, and then share it with your classmates.

Sandy Sim

Using one of Hiroshi's many ingenious inventions, you could design the terrain for a new park. As your hands shape the sand on his digital tabletop, the computer analyzes and displays colors of details such as measurements. As you reshape the sand, the colors change. The colors can show where sunlight and shadows fall, the steepness of slopes, and even where rainwater would flow.

Team up with a classmate. If you were designing the new park, how would you use Hiroshi's invention to

- choose a place for a visitor's center in an area that won't flood?
- create a gentle sloping area for an outdoor theater?
- find the sunniest spot for a swimming pool?

As a class, take turns sharing your design solutions.

JENNA BURRELL

University of California, Berkeley

Twin Passions

Jenna Burrell got an early start in technology—by age eight, she was already programming computers. Growing up to become a programmer seemed an obvious choice—until seventh grade. That's when Jenna's teacher assigned her a report on ancient Mali. "I read everything I could about the African kingdom." Jenna went on to study computer science, but Africa stayed on her mind. Why not mix her two loves? So, off Jenna went to Ghana, in West Africa, to research how the Internet was changing traditional society.

Ghana Get Online

Jenna spent a year in the capital Accra, interviewing people in Internet cafés about how they used the Web. The research sharpened computer-savvy Jenna's new skills as an ethnographer—someone who studies cultures. Unlike Americans she knew, Jenna found young Ghanaians didn't do a lot of reading online when looking for answers to questions. Instead, they preferred asking people in chatrooms. It wasn't just for fun—it was a modern twist on an ancient custom among Ghanaians. "In Ghana, people still ask elders for advice—and they see the Internet as another place to do that," Jenna says.

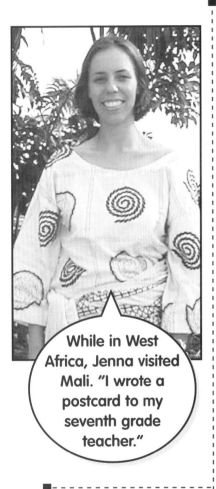

While in West Africa, Jenna visited Mali. "I wrote a postcard to my seventh grade teacher."

Jenna met many young Ghanaians as she researched the Internet's impact on their society.

An ethnographer . . .

studies cultures and societies. Jenna studies how the Internet and other technologies affect people in Africa. Other **ethnographers**

- collect and record fairy tales and folk songs.
- examine how people communicate with jokes.
- study traditional ways of dressing.
- research how people make friends on the Web.

Is It 4 U?

What parts of Jenna's job would you like?

- Working with people from different countries
- Traveling to different parts of the world to learn about other cultures
- Writing about ways in which different cultures meet their needs

Choose one and discuss with a partner what skills you have that would make you a good ethnographer.

Rent-a-Byte

In many places around the world, a person who does not own a computer can rent one at an Internet café. Suppose Kofi, a Ghanaian science student, needs to research a topic on the Internet. In Ghanaian currency, one hour of computer time at the Rent-a-Byte Internet Café costs 2.60 cedis. How many cedis will Kofi pay if he uses the computer for 90 minutes? Round your answer to the nearest whole cedi. How much would this amount be in U.S. currency if one cedi is worth $0.70?

When to Toss a Source

If you used the Internet for research, how would you evaluate the accuracy of your sources? With a group, choose a science topic and then research it. Compare the different Web sites that provide information about your topic. Which sources should you toss and which should you keep? Why? As a class, draw up a science Web Sites Source Directory and list reliable sources that everyone can use.

ANTHONY HORNOF

University of Oregon

"When I started thinking about computers, I wanted to do good and useful things for people."

Seeing Notes

After work, Anthony turns up the volume and deejays at parties. So it's no surprise his next project is software that allows users to make music with only their eyes.

User-Friendly

As a child, Anthony Hornof explored new houses as they were being built. He checked out the mazes of pipes and wires hidden in every wall. They were complicated to put together but easy to use—just flick a switch or twist a faucet. Anthony discovered he loved complex things and how people interacted with them. That eventually hooked him on computers. Studying computer science, Anthony saw how the machines could frustrate people. "I like computers, but I *really* like people," Anthony says. "So I decided to study ways to make computers easier for people to use."

The Eyes Have It

Anthony first learned more about how people used computers. He did so by tracking how people move their eyes as they work on a computer. That got Anthony thinking—could our eyes replace our fingers? He thought in particular about children who have physical disabilities, and who can't move their fingers well. "I wanted to find ways children with disabilities could draw pictures," says Anthony. So he created EyeDraw. It's software that reads eye movements, captured by a video camera, and translates them into computer commands. To draw pictures, all the children have to do is move their eyes!

A human-computer scientist . . .

studies how people and computers interact. Anthony researches new ways people can control computers with their eyes. Other **human-computer scientists**

- ☐ create touch screens that recognize handwriting.
- ☐ design hands-free cell phones.
- ☐ lay out Web sites so they are easier to use.

But Can It Cook?

What do you wish a computer could do that it currently can't? As a team, brainstorm ideas, choose one, and then write a description of the computer program you would develop.

- What task would it perform?
- How would it be helpful?
- How would you communicate with the computer?

Share your ideas with the class.

About You

How have you helped someone do something they couldn't do on their own? In your About Me Journal, describe what you did, how the person felt about your help, and how it made you feel.

Mind Readers?

Advertisers like to observe people as they watch or read their ads. They use computer-linked cameras to record what people's eyes look at first and stare at longest. This reveals what parts of the ad capture their interest.

- Bring in an advertisement that catches your eye.
- Post all the ads in the room. With a partner, study the ads.
- Write down five reasons your eye is drawn to one ad.
- What parts of the ad are you attracted to first? Most? Why?

As a class, discuss what advertisers would learn about your class from using eye-gaze technology. What would they change about their ads?

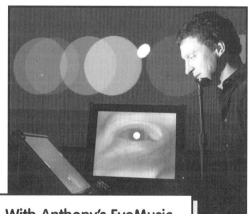

With Anthony's EyeMusic software, you can create music just by looking at the computer screen and moving your eyes.

■ PATRICIA KATOPOL

The University of Iowa

In her free time, Patricia likes to play computer games. "My son always makes fun of me because I play games that always seem to be about food."

Light Change

When Patricia Katopol worked as a lawyer, she saw technology change the way she worked. Tracking down information had meant sifting through mounds of paper and piles of boxes. What a mess! Using computers, lawyers can now find all the same info with a few mouse clicks. "That got me thinking about how people use information," Patricia says. So she changed careers.

Knowledge Know-How

Today, Patricia studies knowledge management. "It's how people obtain, organize, manage, retrieve, and share information," she says. She researches how people handle information at work—in offices, stores, and factories. "Even with technology," Patricia says, "it's still people dealing with information." Imagine a big company where people work at their desks—but still chitchat in the hallways, too. That's a good thing. "Those are the times when people learn a lot," Patricia says. Everyone has information that can be valuable—as long as they can share it. "So it's important that managers understand that learning is going on all the time. One way is when people talk to each other," Patricia says.

An important part of Patricia's research is figuring out how businesses can avoid losing an employee's knowledge when they lose the employee.

A knowledge management expert . . .

studies how to make information useful and accessible. Patricia researches how people get and use information at work. Other **product managers**

- ☐ help surgeons share experiences so they learn from each other.

- ☐ design office layouts that encourage idea sharing.

- ☐ collect advice so they can create guides for students and employees.

Techno Revolution

Most fields have changed because of technology. In medicine, robots assist in surgeries. Scanners let doctors watch organs moving inside the body. And sophisticated internet databases allow scientists around the globe to share their discoveries.

- Select a field such as medicine, communication, transportation, or manufacturing.
- Find out how technology has revolutionized some aspect of that field in the last ten years.
- Identify one or two specific changes. Share your findings with your classmates.

Based on your research, make a prediction about what the next cutting-edge advance might be in that field.

Back in the Day

Back in 1951, a Philadelphia company sold the first commercial computer. Its name was UNIVAC 1, and it could carry out 1,905 operations per second. By 1966, ILLIAC IV was carrying out 300,000,000 operations per second. How many times faster was ILLIAC IV than UNIVAC 1? Round to the nearest whole number.

Compare UNIVAC 1 to today's computers. It weighed 13 metric tons—more than two grown elephants!

Lightning Speed

Today's computers are so fast that new units, called petaflops, had to be invented to measure their speed. One petaflop is equal to *1,000 trillion* calculations per second! The number 1 trillion is written 1,000,000,000,000. Whoa. Use scientific notation to write 1 trillion and 1,000 trillion.

Check out your answers on page 36.

e-Doctor

"Programming was a hobby for me—and all of a sudden it started to become useful to other people." Atul's hospital experience led him to study computer science in college—and then led him to medical school.

ATUL BUTTE

Stanford University

Medical Program

Atul Butte got his start in technology as a young boy when he learned to program his family's home computer. By high school, he was taking college-level computer classes. After school, he volunteered at a New Jersey hospital. There he wrote a computer program that helped pharmacists properly mix intravenous solutions. Today, Atul is a leader in the new, fast-moving field of bioinformatics—using information technology to improve healthcare. Atul knows that one in three kids will grow up overweight and develop diabetes—a serious health condition. But who exactly will it be? Atul is on it!

Patient Download

Atul thinks the answer to who will develop diabetes lies hidden in our genes—the blueprint for life carried in the DNA inside our cells. Atul combines his med and tech smarts to create software programs to sift through the mountains of data on our roughly 30,000 genes. These tools uncover patterns in the electronic data. Atul hopes that one day the data will reveal which genes increase the risk of diabetes. "When we develop a test that will say who will get diabetes, I will be on cloud nine!" Atul says.

Atul is a pediatrician, a doctor who treats children. Here he's treating his daughter to a day at Disneyland.

A physician-scientist . . .

is a medical doctor who treats patients but also spends his or her time doing medical research. Atul creates computer tools that speed up gene discovery. Other **physician-scientists**

- ☐ study how the body fights off infections.
- ☐ research ways of diagnosing diseases.
- ☐ explore the causes of cancer.

The Best of Both Worlds

Atul has always had twin passions—medicine and computers. Pursuing both prepared him for the career he chose.

What are your greatest interests? Do you really enjoy math, but also have a love of art? Are you in tune with music, but also excited by computers and what they can do?

- • Make a list of your interests.
- • Use your imagination to come up with a career that combines two of your passions.
- • Write a brief description of your dream job.

Who knows, you might just find a way to have the best of both worlds.

Computational Teamwork

All over the world, scientists such as Atul put their research data into online databases. This lets scientists share research information quickly and easily.

Scientists submit more than 1,000 files of data each week. At that rate, how many months will it take to upload 26,000 new files? How many files will be uploaded in the next 2¼ years?

About You

What parts of Atul's job interest you?

- • Treating children for illnesses and injuries
- • Using computers to gather and analyze complex data
- • Investigating genetic causes of disease

Write a paragraph in your About Me Journal that explains how your interests would make you a good physician, computer scientist, or both. Why?

Check out your answers on page 36.

> "What I work on every day touches the lives of millions of people looking for better information."

Join the Lineup

When Marissa joined Google, it had 20 employees. Today it has 20,000. "Now there are more people standing in line in front of me for corn on the cob at the company picnic than there were at the entire company when I started."

MARISSA MAYER

Google Inc.

The Right Click

Ever wonder why Google is the way it is? Google it—you might find Marissa Mayer is the answer! Marissa watches over how the world's most popular search engine works—and looks and feels, too. Everything about the Web site is simple, clean, and easy to use. Just type what you're looking for and click. Keeping it simple is a tough job. Marissa works hard to make sure the Googlers she manages do exactly that—keep Google easy to use, even as it grows, and grows.

Google Search, Google Find

Marissa says, "I remember being taught how to use a mouse," and that was in college! That's right. This high-tech guru got her first computer in college. In school, Marissa focused on learning problem-solving skills she'd need as an engineer. First, identify a problem. Next, ask questions. Then, find solutions. When Google was a brand-new, little company, it went searching for engineers to hire. That Google search didn't take long. "Google found *me*!" says Marissa, who became the first female engineer hired by the now giant company.

The simple, clean design of Google's homepage is one of the reasons it's so popular—and powerful.

A product manager . . .

oversees the creation and manufacturing of products for companies. Marissa manages Google's development of new ways to search for information on the Web. Other **product managers**

- ◻ brainstorm ideas for new toys and games.
- ◻ approve TV ads for sports shoes.
- ◻ control improvements to software.
- ◻ supervise introduction of new hybrid car models.

Engineering Design Process

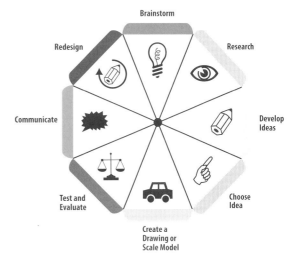

What R U Searching?

It's no surprise that Marissa uses Google to search for information at work. She also uses Google to find snow reports before hitting the ski slopes. What are some ways you use a search engine—for school and for play?

- Make a list of at least five ways and how each is useful.
- As a class, compile all the different uses and the number of students who listed each one.
- What are the five most common uses? What is the percentage of students who listed each?

I'd Like To . . .

Imagine you're on Marissa's team. Is there a new or improved way to communicate or get information that you'd like to develop? With a team, discuss a new information service that could make your life easier, or more fun. Follow the Engineering Design Process above to help you.

- Brainstorm, Research, and Develop ideas.
- Choose one.
- Create a Drawing of your new tool. Include what it is, what it does, and why it will make life easier, or more fun.
- Ask other teams to Test and Evaluate your new tool.
- Use their feedback to guide your Redesign.

About Me

The more you know about yourself, the better you'll be able to plan your future. Start an **About Me Journal** so you can investigate your interests, and scout out your skills and strengths.

Record the date in your journal. Then copy each of the 15 statements below, and write down your responses. Revisit your journal a few times a year to find out how you've changed and grown.

1. *These are things I'd like to do someday.*
   Choose from this list, or create your own.

   - Manage an online community
   - Write new software
   - Create change that makes the world a better place
   - Use computers to turn thought into action
   - Create virtual characters
   - Keep computer information safe
   - Find new uses for computers in health care
   - Study how technology affects people
   - Invent new ways to interact with technology
   - Investigate genetic causes of disease
   - Create new computer products

2. *These would be part of the perfect job.*
   Choose from this list, or create your own.

   - Teaching
   - Building things
   - Writing
   - Testing hypotheses
   - Working on teams
   - Brainstorming new ideas
   - Working in an office
   - Solving problems
   - Learning new things
   - Drawing

3. *These are things that interest me.*
   Here are some of the interests that people in this book had when they were young. They might inspire some ideas for your journal.

   - Working in medicine
   - Making movies
   - Singing
   - Dancing
   - Learning martial arts
   - Exploring the inner workings of computers
   - Programming computers
   - Traveling
   - Playing computer games
   - Volunteering at a hospital
   - Skiing
   - Learning about other cultures

4. *These are my favorite subjects in school.*

5. *These are my favorite places to go on field trips.*

6. *These are things I like to investigate in my free time.*

7. *When I work on teams, I like to do this kind of work.*

8. *When I work alone, I like to do this kind of work.*

9. *These are my strengths—in and out of school.*

10. *These things are important to me—in and out of school.*

11. *These are three activities I like to do.*

12. *These are three activities I don't like to do.*

13. *These are three people I admire.*

14. *If I could invite a special guest to school for the day, this is who I'd choose, and why.*

15. *This is my dream career.*

Careers 4 U!

Information Sciences
Which ˄career is 4 U?

What do you need to do to get there? Do some research and ask some questions. Then, take your ideas about your future—plus inspiration from scientists you've read about—and have a blast mapping out your goals.

On paper or poster board, map your plan. Draw three columns labeled **Middle School, High School,** and **College.** Then draw three rows labeled **Classes, Electives,** and **Other Activities.** Now, fill in your future.

Don't hold back—reach for the stars!

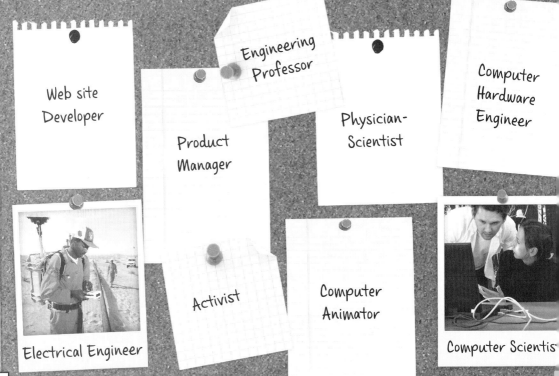

Web site Developer

Engineering Professor

Product Manager

Physician-Scientist

Computer Hardware Engineer

Activist

Computer Animator

Electrical Engineer

Computer Scientist

Computer Architect

Science Illustrator

Digital Librarian

Computer Systems Analyst

Ethnographer

Communications Engineer

Internet Security Analyst

Computer Software Engineer

Manufacturing Engineer

Technical Writer

Computer Game Designer

Human-Computer Interaction Specialist

Electrical Engineer

Computer Theorist

Software Tester

Robotics Engineer

Cryptographer

Glossary

bioinformatics (n.) The collection, classification, storage, and analysis of biochemical and biological information using computers, especially in genetics and genomics. (p. 26)

computer code (n.) A set of instructions for a computer. (p. 12)

diabetes (n.) A disease in which there is too much sugar in the blood. A person with diabetes either cannot make enough insulin, the chemical compound that cells need to take in sugar properly, or cannot use it effectively. (p. 26)

digitize (v.) To convert text, data, or images into digital form. (p. 17)

DNA or deoxyribonucleic acid (n.) A very large molecule that contains all the information for building and controlling a living organism. It is a double-stranded nucleic acid made of nucleotides, with bases adenine, guanine, cytosine, thymine, and a sugar. DNA is found in the nucleus of all cells except in bacteria. (p. 26)

ethics (n.) The discipline dealing with the principles of conduct—what is "right" or "wrong"—governing an individual or a group. (p. 9)

gene (n.) The unit of heredity, encoded as a specific segment of DNA, in living organisms that determines the characteristics that an offspring inherits from its parent or parents. (pp. 26, 27)

genetic (adj.) Describes characteristics that are passed from a parent or parents to offspring. (p. 27)

hybrid (n.) Something, such as a car, that runs on both electricity and gasoline or a power plant that generates electricity from both coal and wind, that has two different ways of performing essentially the same function. (p. 29)

intravenous (adj.) The word intravenous means "within a vein." It usually refers to giving medications or fluids through a needle or tube inserted into a vein, which allows immediate access to the blood stream. (p. 26)

network (n.) A system of computers, computer accessories, and databases connected by communications lines. (p. 13)

petaflop (n.) A petaflop is a measure of a computer's processing speed and can be expressed as a thousand trillion floating point operations per second. FLOPS stands for floating-point operations per second. (p. 25)

pirated (v.) Reproduced without authorization, especially in infringement of copyright. (p. 8)

software (n.) A general term for computer programs and files. (pp. 8, 10, 11, 12, 13, 22, 23, 26, 29)

sound waves (n.) Longitudinal waves that can be heard by the human ear. Sound waves need to travel through a medium, such as air or water. (pp. 10, 11)

technology (n.) The application of scientific knowledge for practical purposes. (pp. 8, 9, 20, 21, 24, 25, 26)

viruses (n.) A computer program that is usually hidden within another seemingly harmless program. A virus produces copies of itself and inserts them into other programs, usually performing a harmful action, such as destroying data. (p. 15)

Index

CHECK OUT YOUR ANSWERS

ACTIVIST, page 7

Phone Home

To get in touch with 30 classmates it will take 4 steps.
Step 1: You call 2 students = 2
Step 2: 2 students each call 2 more students = 4
Step 3: 4 students each call 2 students = 8
Step 4: 8 students each call 2 students =16
4 steps = $(1 \times 2) + (2 \times 2) + (4 \times 2) + (8 \times 2) = 30$ classmates

Night Times

Since each child lives in a different time zone, they all send their text messages at different local times. That means when the text messages arrive in California at 9 P.M. local time, it's 9 P.M. in Oregon, 10 P.M. in Colorado, 11 P.M. in Missouri, and 12 A.M. in Virginia.

- Daniel goes to bed earliest—at 9 P.M.—in Oregon.
- Mia is the second-to-latest to go to bed—11 P.M.—in Missouri.
- If Mia goes to bed earlier than Monica, then Monica must go to bed at 12 A.M.—in Virginia.
- And if Monica goes to bed later than David, then David must go to bed at 10 P.M.—in Colorado.

COMPUTER PROGRAMMER, page 11

Sound Fast?

$1.28 \text{ seconds} = 384,400 \text{ kilometers} \times \dfrac{1 \text{ second}}{300,000 \text{ kilometers}}$

$or\ 238,855 \text{ miles} \times \dfrac{1 \text{ second}}{186,000 \text{ miles}}$

$439.04 \text{ meters} = 1.28 \text{ seconds} \times \dfrac{343 \text{ meters}}{1 \text{ second}}$

$or\ 1,440 \text{ feet} = 1.28 \text{ seconds} \times \dfrac{1,125 \text{ feet}}{1 \text{ second}}$

Flash . . . Bang!

First calculate the distance in kilometers sound travels in 1 second.

$\text{number of kilometers per second} = \dfrac{343 \text{ meters}}{1 \text{ second}} \times \dfrac{1 \text{ kilometer}}{1,000 \text{ meters}}$
$= .34 \text{ kilometers per second}$

$1.36 \text{ kilometers} = 4 \text{ seconds} \times \dfrac{.34 \text{ kilometers}}{1 \text{ second}}$

$.85 \text{ miles} = 4 \text{ seconds} \times \dfrac{1,125 \text{ feet}}{1 \text{ second}} \times \dfrac{1 \text{ mile}}{5,280 \text{ feet}}$

COMPUTER SCIENTIST, page 13

Nibble Kibble

The pet eats
10.5 scoops per week = 1.5 scoops × 7 days
45 scoops in a 30-day month = 1.5 scoops × 30 days
547.5 scoops in a year = 1.5 scoops × 365 days

ETHNOGRAPHER, page 21

Rent-A-Byte

$4 \text{ cedis} = 90 \text{ minutes} \times \dfrac{2.60 \text{ cedis}}{1 \text{ hour}} \times \dfrac{1 \text{ hour}}{60 \text{ minutes}}$

$\$2.80 = 4 \text{ cedis} \times \dfrac{.70 \text{ dollars}}{1 \text{ cedi}}$

KNOWLEDGE MANAGEMENT EXPERT, page 25

Back in the Day

$157,480 \text{ times faster} = \dfrac{300,000,000 \text{ ILLIAC IV operations per second}}{1,905 \text{ UNIVAC 1 operations per second}}$

Lightning Speed

$1 \text{ trillion} = 10^{12}$
$1,000 \text{ trillion} = 10^{15}$

PHYSICIAN-SCIENTIST, page 27

Computational Teamwork

$6.5 \text{ months} = 26,000 \text{ files} \times \dfrac{1 \text{ week}}{1,000 \text{ files}} \times \dfrac{1 \text{ month}}{4 \text{ weeks}}$

$117,000 \text{ files} = 2.25 \text{ years} \times \dfrac{1,000 \text{ files}}{1 \text{ week}} \times \dfrac{52 \text{ weeks}}{1 \text{ year}}$

IMAGE CREDITS

Paul Chesley: Cover. JD Lasica: p. 2 (Verclas), p. 6 top. Seattle University: p. 2 (Quinn), p. 8 top. Webb Chappell: p. 3 (Ishii), p. 18 top. © 2007 Anthony Hornof, used with permission: p. 3 (Hornof), p. 22, p. 23. Sally Ride Science: p. 4, p. 29. NASA: p. 5. Maneno: p. 6 bottom. Carnegie Mellon University: p. 14 top. Kimiko Ryokai: p. 18 bottom. MIT Media Lab: p. 19. U.S. Census Bureau: p. 25. © 2009 Google: p. 28. Clara Lam: p. 30. © WesternGeco: p. 32 bottom left. U.S. Fish and Wildlife Service: p. 33 top left. University of South Florida: p. 33 bottom left. Sandia National Laboratories: p. 33 top right. Dino Vournas: p. 33 bottom right.

Sally Ride Science is committed to minimizing its environmental impact by using ecologically sound practices. Let's all do our part to create a healthier planet.

These pages are printed on paper made with 100% recycled fiber, 50% post-consumer waste, bleached without chlorine, and manufactured using 100% renewable energy.